ABCs
for kids age 1-3
By Dayna Martin

1

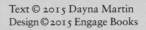

e ENGAGE BOOKS

Mailing address
PO BOX 4608
Main Station Terminal
349 West Georgia Street
Vancouver, BC
Canada, V6B 4A1

www.engagebooks.ca

Written & compiled by: Dayna Martin
Edited & designed by: A.R. Roumanis
Photos supplied by: Shutterstock

FIRST EDITION / FIRST PRINTING

LIBRARY AND ARCHIVES CANADA CATALOGUING IN PUBLICATION

Martin, Dayna, 1983–, author
    ABC animals for kids age 1-3 / written by Dayna Martin ; edited by A.R. Roumanis.

(Engage early readers : children's learning books)
Issued in print and electronic formats.
ISBN 978-1-77226-050-2 (paperback). –
ISBN 978-1-77226-051-9 (bound). –
ISBN 978-1-77226-052-6 (pdf). –
ISBN 978-1-77226-053-3 (epub). –
ISBN 978-1-77226-054-0 (kindle)

1. Animals – Juvenile literature.
2. English language – Alphabet – Juvenile literature.
3. Alphabet books.
I. Roumanis, A. R., editor
II. Title.

QL49.M34 2015   J590   C2015-903399-3
                                     C2015-903400-0

ABCs

for Kids age 1-3
Engage Early Readers
children's Learning Books
by Dayna Martin

ENGAGE BOOKS / VANCOUVER

3

Aa

A baby alligator
is called a hatchling

Alligator

Bb

Bear

A baby bear
is called a cub

5

C c

A baby cat
is called a kitten

Cat

6

Dd

A baby dog
is called a puppy

Dog

7

Ee

A baby elephant
is called a calf

Elephant

8

A baby fox
is called a kit

F f

Fox

9

G g

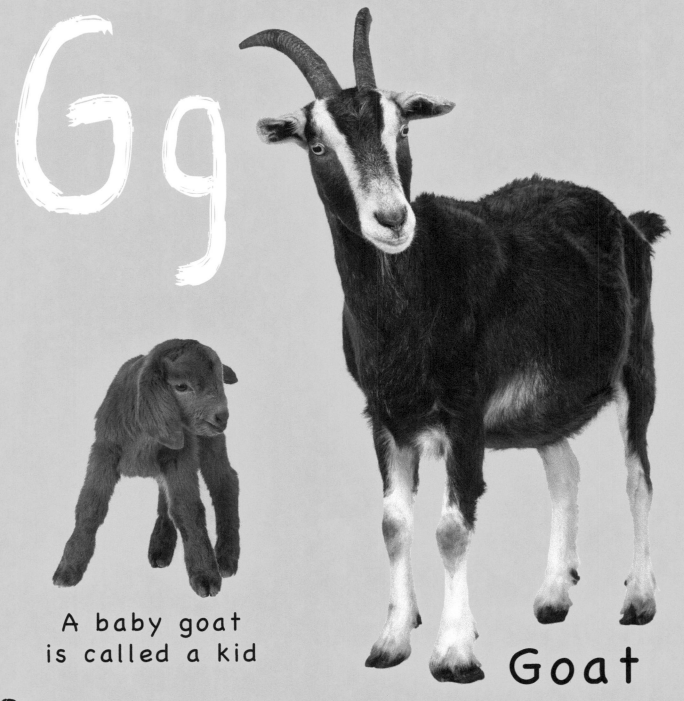

A baby goat
is called a kid

Goat

10

Hh

Horse

A baby horse
is called a foal

11

Ii

Iguana

A baby iguana
is called a hatchling

12

A baby jaguar
is called a cub

Jj

Jaguar

13

Kk

A baby koala
is called a joey

Koala

14

Ll

Lion

A baby lion
is called a cub

15

A baby mouse
is called a pup

Mouse

16

A baby newt
is called a larva

Newt

O o

A baby otter
is called a pup

Otter

18

A baby pig
is called a piglet

Pig

19

Q q

A baby quail
is called a chick

Quail

20

Rr

Rabbit

A baby rabbit
is called a kit

21

Ss

A baby seal
is called a pup

Seal

Tt

A baby tiger
is called a cub

Tiger

23

U u

Uakari

A baby uakari
is called an infant

24

Vv

Vulture

A baby vulture
is called a chick

W w

A baby weasel
is called a kit

Weasel

26

A baby x-ray fish
is called a fry

X-ray fish

27

Yy

A baby yak
is called a calf

Yak

28

Zz

A baby zebra
is called a foal

Zebra

ABCs
activity

Do you remember what these baby animals are called? What three animals are **cubs**, **kits**, or **pups**? Match the baby name to the color below.

Fox

Bear

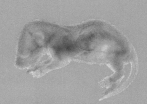

Mouse

Rabbit

Lion

Otter

Weasel

Tiger

Seal

30

Answer: kit

Answer: cub

Answer: pup

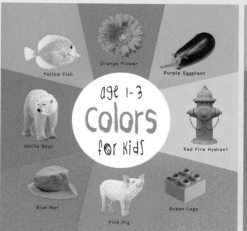

age 1-3
Colors
for Kids

Yellow Fish · Orange Flower · Purple Eggplant · White Bear · Red Fire Hydrant · Blue Hat · Pink Pig · Green Lego

age 1-3
Opposites
for Kids

Out · In · Long · Short · On · Off · Up · Down · Big · Small · Slow · Fast · Old · New · Front · Back

age 1-3
Actions
for Kids

Eat · Jump · Crawl · Brush · Wave · Kick · Swim · Ride

age 1-3
Sizes
for Kids

Small · Medium · Large · Small · Large · Medium · Small · Medium · Large · Medium · Large · Small · Small · Medium · Large · Medium · Small · Medium · Large · Medium · Small · Large

age 1-3
Numbers
for Kids

4 Raspberries · 7 Rubber Ducks · 2 Cars · 8 Presents · 5 Cups · 1 Bowl · 3 Pickles · 6 Balloons

age 1-3
Emotions
for Kids

Angry · Joy · Proud · Shy · Brave · Grumpy · Fear · Shock

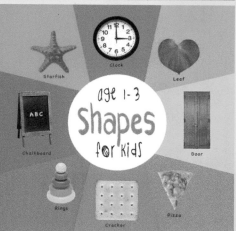

age 1-3
Shapes
for Kids

Starfish · Clock · Leaf · Chalkboard · Door · Rings · Cracker · Pizza

age 1-3
Sounds
for Kids

Ribbit · Moo · Vroom · Flush · Clap · Ring · Roar · Cock-a-doodle-doo

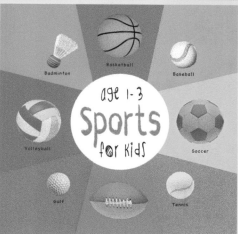

age 1-3
Sports
for Kids

Badminton · Basketball · Baseball · Volleyball · Soccer · Golf · Tennis

CPSIA information can be obtained
at www.ICGtesting.com
Printed in the USA
BVHW022057201021
619395BV00004B/19

9 781772 260502